Airborne

The Search for the Secret of Flight

Richard Maurer

Simon and Schuster Books for Young Readers
Published by Simon & Schuster Inc., New York

In association with WGBH Boston,
producers of NOVA for public television

SIMON AND SCHUSTER
BOOKS FOR YOUNG READERS
Simon & Schuster Building
Rockefeller Center
1230 Avenue of the Americas
New York, New York 10020

SIMON AND SCHUSTER
BOOKS FOR YOUNG READERS
is a trademark of Simon & Schuster Inc.

Manufactured in the United States
of America.

10 9 8 7 6 5 4 3 2 1
10 9 8 7 6 5 4 3 2 1 (pbk.)

Library of Congress
Cataloging-in-Publication Data
Maurer, Richard, 1950
    Airborne: the search for the secret of
flight/Richard Maurer.
    (A NOVABOOK)
    "In association with WGBH Boston,
producers of NOVA for public television."
    Summary: Traces the discovery of the
principles of flight and how this knowledge has been applied to, eventually successful, human attempts to fly.
    1. Flight – Juvenile literature.
[1. Flight.]
I. WGBH (Television station:
Boston, Mass.)
II.Title. III. Series.
TL547.M42 1990
629-13-dc20 90-9710

ISBN 0-671-69422-7
ISBN 0-671-69423-5 (pbk.)

Cover:
**Two pilots and their passen-
ger experience the freedom of
flight in a hang glider.**

Title page:
**Wilbur Wright in a glider over
Kitty Hawk, North Carolina
in 1901.**

To my parents.

At WGBH Boston, Nancy
Lattanzio served as managing
editor; Dennis O'Reilly and
Chris Pullman designed the
book; MJ Walsh served as
typographer; Marianne
Neuman and Virginia Jackson
helped out in various ways.
I thank them, and also Karen
Johnson, director of publishing,
and Paula Apsell, executive
producer of NOVA.
    I am also grateful to illustrator
Brian Lies, copyeditor Ruth T.
Davis, and the staff at the Simon
and Schuster Children's Book
Division.
    John D. Anderson, Jr., Profes-
sor of Aerospace Engineering
at the University of Maryland,
lent his expertise to check the
manuscript. Several libraries
also helped out, notably the
Gale Free Library of Holden,
Massachusetts, the library of the
Museum of Science in Boston,
and Widener Library at Harvard
University.
    The do-it-yourself projects
were created in consultation
with my sons, Sam and Joe.
My wife Susie was, as always,
indispensable.

The NOVA television series is
produced by WGBH Boston.
NOVA is made possible by the
Johnson & Johnson Family of
Companies, Lockheed, and
public television viewers.

Right:
**The ultimate in no-frills flying.
This pilot sits in the open air
guiding an "ultralight" plane
along the coast of North Carolina. Nearby is the beach
where Orville and Wilbur
Wright tested the first airplane almost one hundred
years ago.**

Contents

The Secret

Look up in the sky near a big city for a moment or two and you will spot an airplane – just like that. It's hardly surprising. Airplanes are the normal way of getting around in the air, just as wheels are on the ground, and boats are on the water.

And yet it was not always so. Air travel is a very recent invention. Wheels and boats have been around for thousands of years. But there was a trick to air travel – a secret – that took a long time to discover. Before the secret was known, people who wanted to fly would copy the birds; they would build a set of wings and start flapping. A famous myth tells of the Greek hero Daedalus who built a set of wings and flapped away from his captors. In reality, no human has ever flown this way, although many have tried. It was only when people began to think about what it was they were trying to fly *through* – air – that they started on the road to discovering the secret.

In this book we will retrace that long search. Our plan is to rediscover the principles of flight just as they were uncovered in the course of centuries of patient investigation. We will learn of many false leads, key clues, and dangerous experiments – as we assemble the puzzle to see just what it took to leave the ground behind and, finally, get airborne.

With our agile machines, we can now compete with birds.

Up in the Air

"I believe I have found a way to make a machine lighter than air itself." So wrote Francesco Lana, an Italian priest and amateur scientist, on discovering what he believed to be the secret of flight sometime before 1670.

Lana reasoned that anything less dense than air must float upward like a bubble rising through water. Certainly his theory was valid. Light things do rise through denser ones. A cork bobs to the surface of a pond. Noodles collect at the top of a bowl of soup. Anything lighter than air must ascend through it.

As obvious as this sounds to us, it was not obvious in Lana's time. In order to imagine anything floating in air, people first had to accept that air is a physical substance. A century earlier, the English playwright William Shakespeare had referred to "airy nothing" in one of his plays. The phrase shows how common it was to think of air as having no substance at all.

Air is clearly something. When it moves, things happen. Clouds move, sails billow, trees bend. When the temperature of air changes, rain or snow may fall. Most amazing of all, air sustains life.

When you consider that air is also invisible, these impressive effects give it a magical, almost ghostly quality. It can't be seen, yet it can nourish the lungs, uproot a tree, or freeze a river. No wonder that people once regarded air as beyond understanding.

The view from earth orbit shows that air thins out rapidly into the nothingness of space. Most flight – whether by birds, bugs, or machines – takes place in the realm below the cloud tops (inset), which extends upward to six miles (10 km). Long before anyone could see the atmosphere from space, Torricelli's experiment (far right) proved that an ocean of air exerts pressure on our planet.

◄

Francesco Lana (1631-1687), the first to propose a flying machine based on sound scientific principles.

Weighing the Air

In order for Lana to even *think* that anything could be lighter than air, he had to know that *air has weight*. Thirty years before, in the early 1640s, another Italian scientist named Evangelista Torricelli had proposed an experiment to prove precisely that it does.

Torricelli's idea was to take a long glass tube, sealed at one end, and fill it with mercury. The tube was then to be turned upside down into a bowl filled with more mercury. One might expect all the mercury in the tube to flow out into the bowl. But when the experiment was performed, normal air pressure pushing on the surface of the mercury in the bowl held up the mercury in the tube. The mercury in the tube fell to thirty inches (76 cm) above the surface of the bowl, and then stopped. It takes a *pressure*, or weight, of fifteen pounds per square inch (1 kg/cm^2) to hold up that much mercury.

Interestingly, the height of the mercury not only gives a measure of the pressure exerted by air, it also provides a direct measure of the weight of the *entire* atmosphere. Since air presses on every square inch (or centimeter) of the earth's surface of nearly a billion billion square inches (5 billion billion cm^2), simple multiplication of pressure times area shows that all of the air surrounding the earth weighs roughly fifteen billion billion pounds (5 billion billion kg) – a far cry from "airy nothing."

"We live submerged at the bottom of an ocean of the element air, which by unquestioned experiments is known to have weight," Torricelli proclaimed. This ocean does not crush us because the air that enters through breathing and swallowing, for example, gives us an internal air pressure that equals the air pressure outside of our bodies. The same is true of a fish submerged in water. The fish can survive at any depth as long as its internal pressure is equal to that of the water outside. You might say that *we* are fish in an ocean of air. The main difference is that we are confined to the bottom and cannot swim upward.

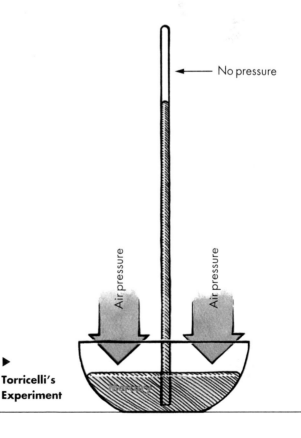

No pressure

Air pressure

Air pressure

▶ **Torricelli's Experiment**

The Mayor of Magdeburg

One of the most dramatic experiments in the history of science was conducted before a large audience in 1657 by Otto von Guericke, an enthusiastic amateur scientist and mayor of Magdeburg, Germany.

After ordering local craftsmen to build two copper bowls about twenty inches (50 cm) across, the mayor joined the bowls at their rims to form a sphere. Through a valve in one of the bowls, he connected a specially-designed pump to remove all of the air from inside. With normal air pressure pressing from the outside, and no air pressure inside, the bowls stuck together.

Then, with a flourish of showmanship, Guericke had two teams of strong horses attempt to pull the bowls apart. Straining against the apparent "nothingness" of air, which was all that was holding the bowls together, the horses were unable to separate them. Guericke then stepped forward and opened the valve, allowing air to reenter the sphere with a hiss. With the pressure now equalized, he could easily pull the bowls apart with his hands.

Francesco Lana in Italy surely must have heard about this stunning demonstration that proved air has weight. He probably counted on using a pump similar to Guericke's to empty the spheres of his proposed flying machine. Yet Lana failed to notice an important point – that once the air is removed, the sphere itself must be sturdy enough to resist the crushing load that presses down on it. Lana's fragile spheres would have crumpled – not flown – under the weight of air.

Two teams of strong horses attempt to pull apart a pair of hemispheres held together by air pressure.

Lana's Airship

Francesco Lana's idea was to break loose from the bottom with the aid of a simple flotation device. This he proposed to make by simply removing the air from the inside of a large hollow sphere. A hollow sphere (a ping-pong ball is a modern example) consists of a rigid shell plus the air contained inside. Suppose the air inside is removed? If the sphere is large enough and the shell thin enough, removing the air will reduce the sphere's weight to less than that of the surrounding air, and it will rise.

Lana calculated that a thin sheet of copper made into a sphere fourteen feet (4.3 m) in diameter would weigh 1,848 ounces (52 kg). According to his estimate, the air inside the sphere would weigh another 2,156 ounces (61 kg). If that air were subtracted, the entire sphere would be lighter than an equal amount of the air surrounding it. And up it would go.

By increasing the size of the sphere four times, Lana figured that he could build a device capable of lifting a couple of people into the air. Four such spheres strapped together would make a practical airship, complete with sails, oars, and a passenger compartment shaped like a canoe. "I have conferred on these matters with many sage and well-instructed persons who have not been able to find any error in my discourses," Lana wrote confidently.

Alas, neither Lana nor his advisors realized that if air was removed from his spheres, the inside pressure would no longer equal the outside pressure. The structure would collapse under the weight of the atmosphere, like a tent with the supports knocked away. A properly reinforced sphere, able to keep its shape with all the air removed, would be too heavy to ever get off the ground. Fortunately, Lana never had to experience this disaster since he didn't have enough money to build his machine.

Even though it wouldn't have flown, his invention was an important advance. Until this time would-be flyers had invariably imitated the birds. They had ignored the actual properties of air, which will not support a dense thing like a human waving a couple of boards. By carefully considering the substance in which flight is to take place (air), Lana came up with a completely fresh and scientific approach. Although it didn't work, it at least promoted the idea that the problem could be solved by serious thought.

Other scientists soon discovered the error of Lana's reasoning, but it would be many years before the next step was taken.

Lana's airship. The man in the robe is apparently Lana himself, explaining the secret of his flying machine.

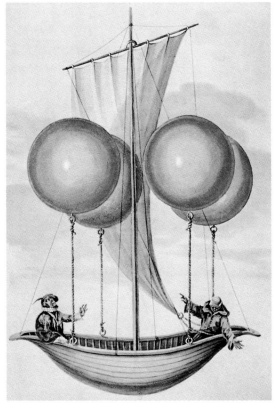

Tower Jumpers

Birds are such excellent flyers that they make flying look all too easy, tempting some humans to fashion a pair of wings and try it themselves. The usual test flight took place from the highest spot available, a tower. Tower jumping became quite common in the fifteenth and sixteenth centuries. The only explanation for its popularity is that these often fatal attempts were encouraged by the reported successes of others.

There was, for example, the Italian scholar who made several training flights over a lake before putting on a public demonstration in which he flew across a city square using a rowing device with wings. At least, so it was said. But this was nothing compared to the Turk who supposedly flew great distances, "turning round and round in the air," with wings like those of an eagle.

The truth, if any, to such tales probably comes at the end of the story, which often reports the death or crippling of the pilot due to some misfortune in flight.

A tower jumper prepares to test his wings.

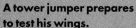

The Secret of Hot Air

One beautiful late autumn day in 1783, 113 years after Lana had published the idea for his flying machine, a bright blue sphere – forty-six feet (14 m) across – rose from a plaza near Paris, France, lifting a circular platform holding two passengers. It was a balloon ascent – the first to take humans on a flight through the air.

How was it possible for a machine so similar to Lana's to succeed where his could not? What was the mechanism that made this craft rise?

While Lana's strategy had been to make a machine lighter than air by *removing* the air, the Montgolfier brothers of France, builders of the blue balloon, relied on air – but this time *hot* air.

The flight proved that hot air creates the same effect that Lana was trying to achieve with no air. Hot air is less dense than normal air; in other words, it's lighter. When you light a candle, the flame heats the surrounding air and causes it to rise. The warmth that you feel when you hold your hand above the flame is hot air rising. Fill a lightweight container with enough hot air and it, too, will rise.

Joseph and Etienne de Montgolfier made this discovery without resorting to any theory. They simply noticed that smoke rises, and wondered if it could be used as a lifting agent. When they tested this idea by filling a large silk bag with smoke, it worked. The bag rose. Because the brothers thought that the smoke itself was doing the lifting, they made the smokiest fires possible for the rest of their experiments. (As was later learned, smoke itself is not necessary; it's the heat that's important.)

How much of a lift can you get from hot air? It depends on how much you use. A container of hot air (at a temperature somewhat hotter than boiling water) weighs only three-quarters as much as the same size container of room-temperature air. The hot-air container will want to rise through room-temperature air because it is less dense.

Heated air rises from a candle flame.

▶
The Montgolfier brothers test one of their first hot-air balloons. The craft was made with strips of cloth fastened by 1,800 buttons.

Suppose, for example, that you fill a large balloon with 10 pounds (4.5 kg) of room-temperature air. (This much air fills a space about the size of a small closet.) A balloon filled to the same size with hot air will weigh only 7½ pounds (3.4 kg) – three-quarters as much. The difference in weight – in this case 2½ pounds (1.1 kg) – is the *lifting force* .

Now, if the material that makes up your balloon weighs as much (or more) than the total lifting force, there will be no force left over to raise the balloon. But suppose that you use just 1 pound (0.4 kg) of a light-weight material – like rubber or silk – to construct the balloon. You will still have 1½ pounds (0.7 kg) of lifting force left over. That's not much, but it's enough to lift something like a small kitten.

To produce enough lifting force to lift a human, you will want a balloon at least 100 times larger. The Montgolfier brothers chose a balloon *500 times* larger for their first passenger balloon. Naturally, they wanted to go themselves, but their father would not allow it. The King of France suggested that condemned criminals be sent. But a couple of brave Frenchmen (who were aware of the honor attached to being the first to fly) eagerly volunteered and were accepted. They were a physicist, Jean-Francois Pilatre de Rozier, and an army major, the Marquis d'Arlandes.

Seven and one-half pounds (3.4 kg) of hot air can lift two and one-half pounds (1.1 kg) through air at room temperature. Ten pounds (4.5 kg) of room-temperature air, which fills the same amount of space, goes nowhere.

On November 21, 1783, everything was ready. A test flight had already shown that ballooning was reasonably safe, since the test subjects – a sheep, a rooster, and a duck – had all returned alive. Pilatre and d'Arlandes climbed into the partly enclosed platform suspended beneath the seven-story-high balloon. They topped off their hot-air supply with a brief blaze of shredded wool and straw (producing plenty of unnecessary smoke). After a false start that almost set their craft ablaze, the courageous airmen took to the skies.

Two Frenchmen become the first to fly, piloting a gaily decorated hot-air balloon built by the Montgolfier brothers.

One witness was Benjamin Franklin, who had come to town for the peace conference that ended the American Revolution. "When it went over our heads," he wrote to a friend, "we could see the Fire which was very considerable." The Montgolfier brothers had rigged a burner beneath the open base of the balloon to maintain the hot air while in flight. Without it, the air inside the balloon would quickly cool and the craft would descend. "When they were as high as they chose to be," Franklin noted, "they made less Flame and suffered the Machine to drive Horizontally with the Wind."

It flew twenty-five minutes before setting down in the country, to the astonishment of the local people.

Build a Hot-Air Balloon

Relive history by discovering the lifting force of hot air. (No smoke needed.)

Materials
- Five pieces of tissue paper (Note: The tissue paper sizes shown in the diagram are based on the standard size sold in party and stationery stores. If you have larger sheets, use them: The bigger the balloon, the better it works.)
- Scissors
- Glue
- Hair Dryer

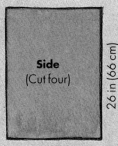

Side
(Cut four)

20 in (51 cm)

26 in (66 cm)

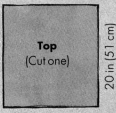

Top
(Cut one)

20 in (51 cm)

20 in (51 cm)

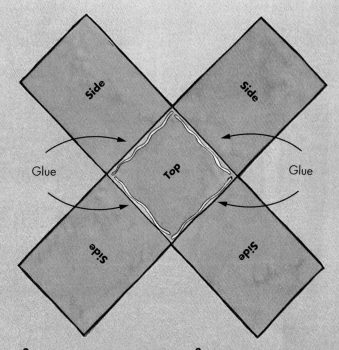

Side

Side

Top

Side

Side

Glue

Glue

1
Cut out four sides and a top.

2
Glue sides to top as shown (above). Then fold the sides up and glue the long edges together to make an open-ended bag (below).

3
Make the opening smaller by gluing the corners together.

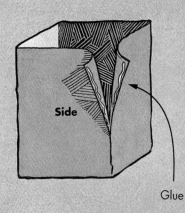

Lift

Side

Glue

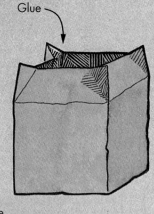

Glue

4
Fly your balloon: Your hot-air source will be a hair dryer. Choose the hottest setting, insert the dryer through the opening of the balloon, and warm the air inside for about one minute. Switch off the dryer. The balloon should float upward. (Note: This experiment works best in a cold room.)

Inflammable Air

A few months earlier, the French physicist J.-A.-C. Charles had heard about the Montgolfiers' experiments. Not knowing the details, he guessed that the brothers had used a recently discovered gas that was only one-fifteenth as dense as air. The new gas was called "inflammable air" because it was highly explosive. Today, we know it as hydrogen, the lightest of all elements. Eager to get aloft himself, Charles hastened to construct what he thought was an identical machine.

Hydrogen is so light that any of it released into the air eventually rises into space. So how does one collect it? Hydrogen can be found inside many complex substances such as water. Nowadays, hydrogen is usually produced by putting an electric current through water. The energy breaks up the water into hydrogen and oxygen gas. Since no one in Charles' time knew of this technique (nor did they have access to enough electricity), he used a more dangerous method in which a very corrosive chemical – sulphuric acid – is mixed with iron filings. The vigorous froth that results is hydrogen gas escaping from the acid.

The Montgolfier brothers had done nothing quite so dangerous. They simply made a fire and collected the hot air. However, a hot-air balloon is quite risky since it actually has a fire burning beneath it. Because of this, Pilatre and d'Arlandes spent much of their historic flight worrying about going down in flames. Indeed, one of d'Arlandes' duties during the mission was to extinguish the smoldering patches of balloon fabric that had caught fire from stray sparks.

Though hydrogen itself is inflammable, if care is taken to avoid sparks and fire, it makes an ideal gas for balloons. Hydrogen has almost four times the lifting power of hot air, and, unlike hot air, it does not need replenishment in flight. When Charles built his own two-passenger balloon, the craft was much smaller than the comparable Montgolfier machine.

All in all, the hydrogen balloon was a much safer bubble on which to ride up through the air.

Inflating the first hydrogen balloon. Iron filings and sulfuric acid were combined in the barrel to produce hydrogen gas.

A flame flickering through the open base of a modern hot-air balloon heats the air inside to make the craft rise.

Up, Up, and Away

The excitement of ballooning has been revived in recent years thanks to the combination of lightweight parachute fabric, fire retardant chemicals, and powerful propane burners. These allow a much safer and more versatile hot-air balloon than was available to the Montgolfier brothers in the late 1700s.

Hot-air trips unfortunately last only as long as the fuel supply, which is used to reheat the air as it cools. For sustained travel, however, nothing beats gas – either hydrogen, or denser, non-flammable helium. Once filled, gas balloons can stay up for days. They could stay up forever; but, because they expand as they rise (due to the drop in atmospheric pressure with altitude), they must slowly release gas to prevent over-expansion and a burst balloon.

Gas Balloon Record, 1961 ● **20**
(32)

● **15**
(24)

Hot-Air Balloon Record, 1988

● **10**
(16)

A hot-air balloon in flight.

A gas balloon, rising high over the Alps. Note the difference in shape from the hot-air variety, shown here at the same scale.

● **5**
(8)

Maximum Altitude of Goodyear Blimp, 1990

Charles Gas Balloon, 1783
Montgolfier Hot-Air Balloon, 1783

Altitude in miles (km)

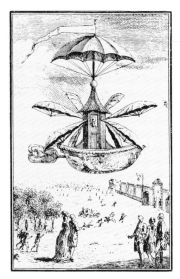

Eager to fly by any means, French inventor Jean-Pierre Blanchard worked on flapping contraptions like this one (which he never tested) before finally taking up ballooning.

The World Aloft

The relative safety of his invention gave Charles the opportunity to enjoy the unusual experience of being airborne. Reporting on the first passenger flight in his craft, which Charles himself took with a colleague just a few days after the flight of Pilatre and d'Arlandes, he found "all was glorious – a cloudless sky above, a most delicious view around."

"Oh, my friend," he told his companion, "how great is our good fortune! I care not what may be the condition of the earth; it is the sky that is for me now."

Flight enthusiasts everywhere agreed, abandoning their strange flapping contraptions to take up ballooning. One convert was Jean-Pierre Blanchard, a Parisian inventor who had been working on aircraft designs that featured an enclosed tub with a system of revolving sails. Impressed by the spectacular successes of the Montgolfiers and Charles, Blanchard junked his machine, declaring it as incapable of flight as "the heavy ostrich."

Soon Blanchard was setting records in balloons. During a career of almost sixty lighter-than-air trips, he was the first to pilot a craft across the English Channel, the first to test a propeller in flight, and the first to release parachutes from on high (sending down animals). He was also the first to demonstrate balloons in Germany, Holland, Belgium, Switzerland, Poland, and the United States, where he made an ascent witnessed by George Washington.

People everywhere were seeing with their own eyes that humans could fly at last – perhaps not with the freedom of birds, but who could argue that it was not really flight? All that was needed, announced Franklin, was a means of guiding the balloon where the pilot wanted to go, rather than where the wind happened to blow.

▶

Hot-air balloons float downwind over a glorious landscape in New Mexico. One of the first to describe the view from aloft was J.-A.-C. Charles (inset), who with a friend took the first ride in a hydrogen balloon on December 1, 1783.

Winging It

One flight enthusiast refused to give up on wings – a young Englishman named Sir George Cayley. Cayley realized that there is an important difference between floating – what a balloon does – and flying – what a bird does. Cayley wanted to fly.

In some ways the secret of winged flight is as obvious as the fact that smoke rises. Cayley happened upon this discovery in 1796, when he constructed a popular toy of the day. Made of feathers, two corks, a stick, some string, and whalebone, it is what we would call a helicopter. The feathers acted like helicopter blades; a bowed piece of whalebone with a tight string acted like a spring to make them rotate. People were amused to watch the little craft rise to the ceiling. Cayley was more than amused. He was intrigued. Though he did not understand the principle that made

the toy fly, he was aware that something about the shape and position of the feathers was crucial. He also knew that not just any feathers would work. They had to be the stiff, arched feathers from a bird's wing.

While he was still in his early twenties, Cayley determined to follow up this personal discovery of something that everyone knows. Feathers can fly.

Sir George Cayley (1773-1857), pioneer investigator of wings and how they work.

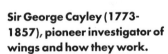 **Cayley's toy helicopter. The tension of the bent piece of whalebone caused the string to unwind, turning the feathers.**

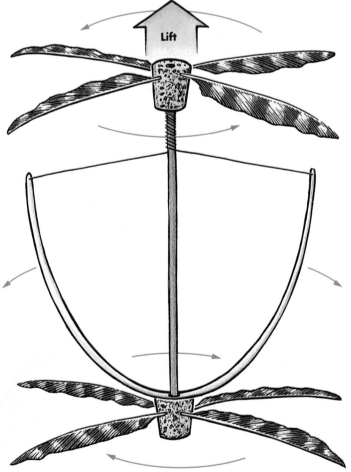

Getting a Lift

If you look closely at a bird feather, you can see stiff filmy vanes that are attached to a long pointed shaft. The vanes on one side of the shaft are longer than on the other. Look at the feather from the end, and you will see that the vanes form a gentle arch, like a rainbow. Sweep the feather through the air, holding it with the arch rightside up and leading with the short vanes. The feather will pull upward. This is its magic.

In Cayley's helicopter, eight such feathers swept through the air, arch up. A French version of the toy used silk-covered wire frames with a twist instead of an arch. Both toys flew. A windmill and a sail work on the same principle. Both present a tilted or arched surface to the wind and generate a force that either turns rotor blades or pushes a boat. Windmills and sailboats had been in existence for many centuries, but before Cayley no one had given much thought to harnessing the same force for flight.

Hold the feather with the short vanes leading, sweep it through the air, and you will feel a slight upward pull – the force of flight.

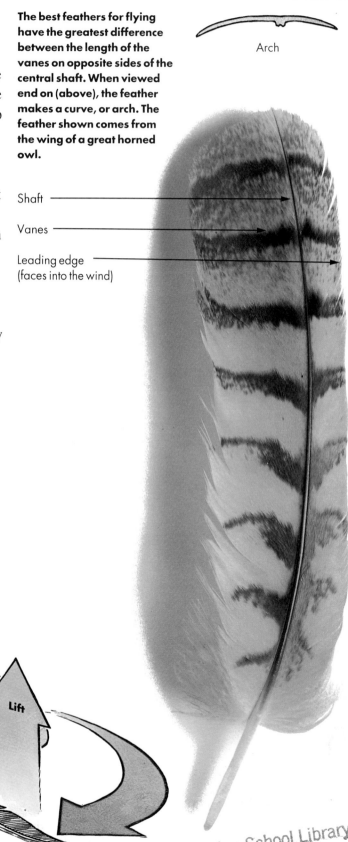

The best feathers for flying have the greatest difference between the length of the vanes on opposite sides of the central shaft. When viewed end on (above), the feather makes a curve, or arch. The feather shown comes from the wing of a great horned owl.

Arch

Shaft

Vanes

Leading edge
(faces into the wind)

Lift

Seed-copters

Maple seeds have a single wing – not to travel up, but to twirl down, carried by the autumn wind to a potential spot for a new maple tree.

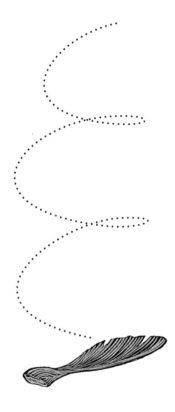

"Q"-copter

Cayley's toy helicopter was made with common materials of his day, including feathers and a bit of whalebone. You can build a similar machine with cardboard and a Q-tip.

Materials

- Cardboard (such as that on the back of a pad of paper)
- One Q-tip with a wooden shaft
- Spool
- Tape
- String
- Scissors

1

Trace propeller pattern onto cardboard. Cut out along solid lines.

2

Cut slits on blades where marked. Then make a groove along dotted lines with a ball-point pen. Fold blade edges down slightly. Now fold each entire blade up slightly.

3

Punch a small hole in the center of the propeller. Remove one swab from Q-tip and insert through hole on the side that has blades bent up. Push the Q-tip through till the swab on the other end fits snugly in the hole. Wrap a small piece of tape around the Q-tip just below the propeller.

4

Wind string counter-clockwise around the Q-tip. Start from the bottom and work toward the top. Leave enough string to hold onto.

5

Insert Q-copter in spool. Tape bottom hole closed. Give the string a firm, steady pull. Q-copter should fly up to the ceiling.

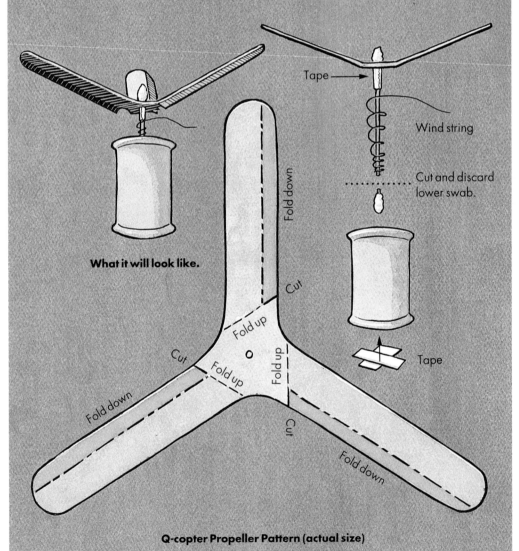

What it will look like.

Tape

Wind string

Cut and discard lower swab.

Tape

Fold down

Fold up

Fold up

Fold up

Cut

Cut

Fold down

Fold down

Cut

Q-copter Propeller Pattern (actual size)

Where does this force come from?

If you think about a board held in a strong wind with the front edge tilted slightly up, it should be clear that the bottom of the board will catch some of the breeze and get an upward push. But a much stronger force is generated by another effect: suction. Suction often occurs where there is a pressure difference. Sucking on a straw, for example, reduces the air pressure inside the straw. The normal air pressure on the liquid outside causes the liquid to flow up into your mouth.

A board tilting into the wind also experiences suction. Part of the airstream strikes the board on the bottom and gets deflected down and away (giving the board a bit of an upward push). The air itself follows a direct course and keeps a fairly constant speed. Air travelling along the top of the board takes a more curved path, and speeds up. Because faster-moving air has lower pressure than slower-moving air, the board is "sucked" upward – just as low air pressure in a straw will suck up liquid.

You can experiment with this force by putting your hand outside the window of a moving car (not too far). With your hand flattened and your fingers together, point your fingers in the direction the car is moving; now tilt your hand up just a bit. You will feel a strong upward pull. Next, try keeping your hand level, but arch it slightly. You will again feel an upward pull. In this second case, the airflow has a longer distance to travel over your hand because of the arch – with exactly the same result as for a flat shape held at an angle. In fact, the arch works even better than a flat shape, because it can produce suction when level, and even more suction when tilted. This is the shape that most modern airplane wings have.

Three examples of suction caused by a pressure difference. (Left) Sucking on a straw lowers air pressure in the straw; the higher air pressure in the glass forces liquid into your mouth.

Equal pressure

Lower pressure
Higher pressure

(Top right) Air rushing past a tilted object travels faster along the top than along the bottom; the result is lower pressure on the top and a suction force pulling the object upward. (Bottom right) An arched shape creates even more suction by this method.

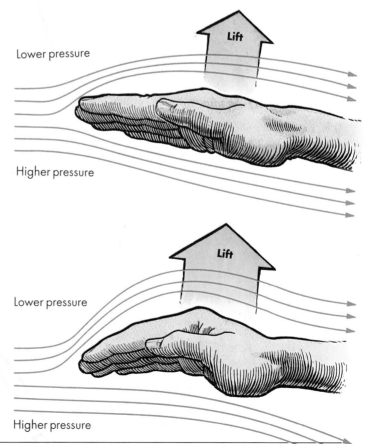

Lift

Lower pressure

Higher pressure

Lift

Lower pressure

Higher pressure

Lift and Thrust

One of the earliest experiments Cayley tried was measuring the amount of lift produced by flat shapes tilted at different angles. He then used this knowledge to build a model glider, which consisted of a long pole, a fixed (unmovable) wing, and a cross-shaped tail. To understand what a breakthrough this simple device was, we need only look at another of the proposed flying machines of the day.

Thomas Walker's "flapper" is a typical example of one. The English portrait painter designed (but never built) a craft that was little more than a giant bird suit. The pilot was supposed to climb inside and move a set of levers to make the wings flap. Walker obviously theorized that something that looked like a bird, and acted like a bird, ought to fly like a bird.

Walker was not alone in his thinking. Flapping was a key feature of almost all winged flying machines before Cayley's discovery. None ever flew, but geniuses such as the artist and inventor Leonardo da Vinci labored diligently on them, devising ever more clever ways of working the wings.

With birds, bats, and insects serving as the only working examples, it's easy to see why people believed that flapping is the sole cause of flight. Cayley's experiments with his toy helicopter taught him otherwise. It flew merely by turning feathers horizontally through the air. Air rushing past feathers was enough to provide lift. He realized that the toy explained one of the most puzzling feats of the birds – that they can glide for long distances without moving their wings.

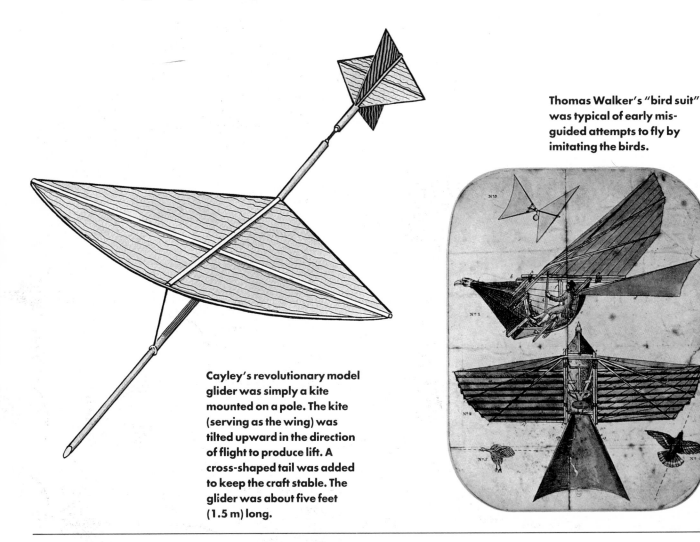

Thomas Walker's "bird suit" was typical of early misguided attempts to fly by imitating the birds.

Cayley's revolutionary model glider was simply a kite mounted on a pole. The kite (serving as the wing) was tilted upward in the direction of flight to produce lift. A cross-shaped tail was added to keep the craft stable. The glider was about five feet (1.5 m) long.

Ode to Wings

You cannot fly like an eagle with the wings of a wren.
– William Henry Hudson
English author, 1909

Or vice versa. Eagles are built for soaring; wrens for darting. One has long, broad wings for efficient gliding; the other short, rounded wings for maneuvering in tight spaces. Animals and airplanes need different wings for their different purposes. Two wings, for example, serve the biplane for low-speed flying. Four wings, moving independently, allow the dragonfly to dart in all directions. A machine like the Stealth bomber is *all* wing – a shape chosen for its ability to slip through radar unseen.

Wings also have different shapes when viewed end-on: (top to bottom) a typical insect wing; a bird wing; a passenger jet wing.

Paraglider

Bald eagle

Passenger jet

Home-built sports plane

Dragonfly

Two-seat biplane

Stealth bomber

So why flap at all? Cayley reasoned that birds flap for *propulsion* – to get themselves going and to keep going, much as a locomotive gets a train rolling and keeps it up to speed. A bird can glide downward without flapping, just as a train can roll downhill without power. But to get back up again, both need propulsion, or *thrust* . (Soaring birds can get higher without flapping, with the aid of a brisk wind or a rising current of air. Nature provides the propulsion in the form of moving air.)

By figuring out the dual purpose of a bird's wing – to supply *both* lift and thrust – Cayley came up with a revolutionary idea. Why not separate the two? Use a fixed wing for lift, and devise a separate system to get the craft moving. Cayley's glider was a brilliantly successful test of the first part of this idea.

The second part – inventing a separate propulsion system – caused him more difficulty.

The Big Engine that Couldn't

Centuries before the Montgolfier brothers lit a fire beneath a small silk container and discovered the hot-air balloon, people in China were celebrating festivals with exactly the same device. But the Chinese never made huge passenger-carrying versions.

Indeed, any civilization with access to finely woven cloth and varnish to make it airtight could have been airborne, probably as early as the time of the ancient Egyptians 5,000 years ago. Making a balloon is that simple. Cayley's model glider is not so difficult, either. Anybody from the Stone Age on could have built one, and maybe someone did. The native people of Australia built something that flew even better – the boomerang.

The technology of propulsion is a different matter, however.

Different proposals for propelling balloons: using sails combined with explosions (below); flocks of eagles (right); and paddles (far right).

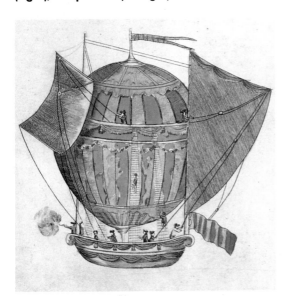

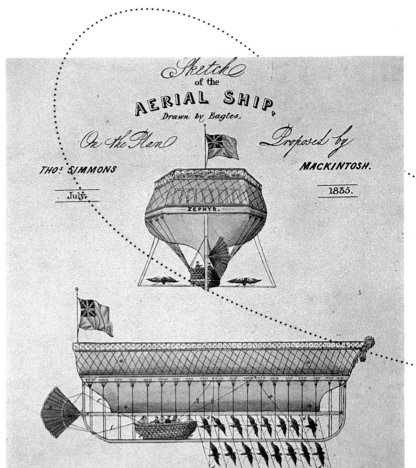

Propulsion had also been causing problems for balloonists. Benjamin Franklin and others believed that it would be a simple matter to figure out a method to move balloons against the wind. After all, boats had no trouble travelling against a water current. Oars and sails worked quite well.

Therefore, oars and sails were tested on balloons. The record-setting Jean-Pierre Blanchard was at the forefront of these attempts. He tried huge double oars in the shape of webbed feet. No luck. He unfurled sails from the passenger basket. No effect. His colleagues suggested using paddle wheels, jets of hot air, explosions of gunpowder, and even hitching silk cords to eagles. Whatever was tried, there seemed no way of making headway through the ocean of air. The balloon is too bulky an object to push through the air without using substantial force – more force than can be supplied with the kinds of devices tested by Blanchard.

Working independently, Cayley suggested streamlining the spherical balloon into a cigar shape, like a boat, which would more easily penetrate the resisting breezes. As with his fixed-wing idea, Cayley had again come up with part of the solution to the problem of efficient flight – but only part.

Return to Sender

One of the simplest yet most sophisticated flying machines is the boomerang, which combines elements of a wing and a spinning top to produce loops, figure eights, spirals, and other tricks of flight. The theory behind the motion of a boomerang is quite complicated, though this hardly mattered to the clever Australian natives who perfected it as a weapon. They cared only that the boomerang returned for quick reuse when it missed its prey.

You can make a miniature boomerang for indoor use by copying the pattern below. To launch it, press the "v" of the weapon just under your thumbnail, hold it horizontal, and give it a swift flick.

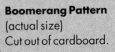

Boomerang Pattern
(actual size)
Cut out of cardboard.

III.ᵐᵉ EXPERIENCE DE Mʳ BLANCHARD

In order to propel either a balloon or a winged aircraft, a power-producing machine is needed. In 1773, the year Cayley was born, the chief power-producing machine in England was the waterwheel. But water power hardly qualified for use in the air. That same year, however, saw a significant symbol of change. The Scottish inventor James Watt proposed that his new steam-powered engine be rated in "horse-power," or the number of horses that could do an equal amount of work. One of the first of Watt's revolutionary engines was rated at ten horsepower.

The miracle of the steam engine was that it operated on wood or coal, rather than on flowing water. The result was a convenient power source that could be placed almost anywhere. Unfortunately, the steam engine was a massive object, weighing about as much as the horses it was designed to replace.

By the time Cayley was grown up, the steam engine had been improved and made light enough so that it could be equipped with iron wheels to pull a train of cars along a steel track. Thus the train was invented. Unfortunately, the train engine was still far too heavy for use in a flying machine.

Cayley's Calculations

Cayley's experiments had shown that the lift produced by a wing is determined by its size, and also by its angle of tilt and speed. He had determined that one square foot (0.1 m^2) of wing surface, tilted upward at an angle of six degrees, and moving through the air at twenty-five miles per hour (40 km/h), could lift one pound (0.5 kg) into the air. A generous wing of 400 square feet (40 m^2) – about the area of nine ping-pong tables – could therefore support 400 pounds (200 kg). This was about as large a wing as Cayley was willing to consider, and it had to be able to lift plane, pilot, and power source.

The speed Cayley chose may seem rather slow today. However, he had observed birds gliding at this velocity, and it was still a faster speed than trains of his day could go. Cayley knew that higher speeds meant greater lift, and he predicted that one day flying machines would reach an astonishing 100 miles per hour (160 km/h). Nowadays, some jet planes can travel twenty times faster.

One of James Watt's advanced steam engines at work in the 1790s.

According to Cayley's estimates, a two-horse-power engine (which is about equal to a small power lawnmower today) would produce just enough power to lift a plane and pilot off the ground. Since the motor would also have to lift itself, it had to be very light – no heavier than the pilot, Cayley calculated. Steam power would probably never do, he realized, since it required heavy tanks for boiling and condensing water. An entirely new type of engine was needed.

Cayley noticed something else in his figures. Because humans can generate two horsepower for very brief periods – while running up stairs, for example – it ought to be possible for a person using every muscle available to produce enough power to make a plane fly.

And so, Cayley reached two conclusions that no one else had worked out: in order to propel a flying machine, an engine would have to be vastly more efficient than any in existence; and human-powered flight was possible, but only with the utmost physical exertion.

How Humans Fly

Whether they knew it or not, "flappers" and "tower jumpers" were really aiming at a craft like the one below. Weighing a mere sixty-eight pounds (31 kg), the pedal plane *Daedalus 88* is heavier than many of the machines of old, yet it is capable of long-distance flight under human power.

The two-horsepower output of a vigorously-exercising athlete is all that's needed to pedal *Daedalus 88* off the ground and into the sky. Once aloft, this and similar human-powered planes require about one-half horsepower to keep them cruising. The secret, as Sir George Cayley predicted, is the combination of a high-lift, low-speed wing (just like that of the heron, his favorite bird) and an efficient propulsion system. Propulsion, in this case, is supplied by a propeller powered by the strongest muscles in the human body – those of the legs. Space-age composite materials are crucial to keeping down the weight of the vehicle, which explains why humans didn't fly under their own power until, well, the space age.

By making a human-powered air voyage from the Greek island of Crete to the nearest land, *Daedalus 88* accomplished in reality what the ancient Greek hero Daedalus did in legend. Note the pilot's legs working the pedals, which in turn drive the propeller.

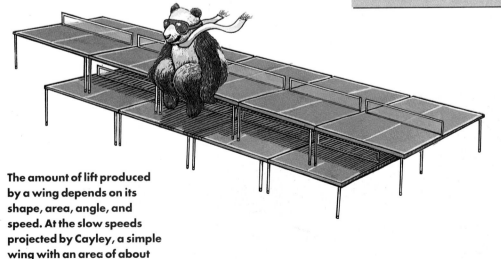

The amount of lift produced by a wing depends on its shape, area, angle, and speed. At the slow speeds projected by Cayley, a simple wing with an area of about nine ping-pong tables would have been able to lift a 400-pound (200-kg) object.

Down, Twist, and Around

Cayley then set out to invent a new engine and to discover how humans might fly under their own power. Both problems required that he first discover how to propel an aircraft efficiently, whether the power source was motor or muscle. It was the same hurdle faced by the balloonists. How do you get a grip on the ocean of air?

Having gotten his inspiration about lift from feathers, Cayley next took a close look at flapping to learn something of propulsion. Birds flap so quickly that it's hard to follow their motion in detail. The most logical explanation scientists had proposed was that wings acted like boat paddles in water, pushing air backward in order to move forward. Leonardo da Vinci and other early investigators had believed that wings beat backward during forward flight, and downward when the bird was going up. Cayley believed that something more complicated was going on.

There are nearly as many bones in a bird's wing as in the human arm and hand. Feathers are attached along the whole length of the wing, but those that emerge from the hand-like structure on the end are the most flight-efficient. These feathers – called the "primaries" – are the strongest and have the best shape for creating lift. They are also the most versatile, since the "hand" can twist them to practically any position.

After observing where the muscles attach to the bones, Cayley theorized that the basic motion of the wing is up and down, and that the end of the wing twists during each beat to move the air backwards. You can copy this motion by flapping your arms up and down; on the downstroke twist your hands to push air down and back. (If your hands were bigger, and your arms stronger, you could probably lift yourself off the ground this way.)

Primary flight feathers are best for propulsion.

Secondary flight feathers make the wing into a curved shape for high lift.

A bird flies by moving its wings in a cycle: down and forward, then up and backward.

How Birds Fly

This photograph of royal terns and gulls in flight shows the different stages of the wing-beat cycle. Notice how some of the birds have their wings extended stiffly down and forward. This is the powerful downstroke that provides both lift and propulsion. The feathers on the inside of the wing – or "arm" – supply an upward-directed force during flight; those on the outside – or "hand" – splay outward to provide a forward force.

On the upstroke, the wings sweep back and up while folding almost in half. Human swimmers do the same, holding their arms stiff during the power stroke, and flexing them during the return. When the wings almost clap together above the bird, the feathered flyer is ready for the next downstroke.

These birds are in the process of taking off – a maneuver that requires a lot of power. Once they gain altitude and speed, the extreme upward and downward sweep of the wings will become less exaggerated.

Cayley tried to imitate the twisting motion of the outer wing by developing a special paddle with hinged slats arranged like the bird's primary feathers. Two paddles were attached to either side of his glider like oars, and were flapped up and down. On the downstroke, the end of the paddle would automatically twist to push the air down and back. On the upstroke, the slats would hinge open so that the air would pass freely through them. Cayley admitted it was "a complicated action depending upon a nice adjustment of all the parts." Down, twist, and back up; down, twist, and back up. That was the basic motion.

Working the paddles vigorously, Cayley's assistants could get a small amount of thrust, but never enough to get off the ground. After so many breakthroughs, he had finally come up against a dead end. Cayley never realized how close he was to the solution.

Down, twist, and back up. If you had stood behind the glider and watched it struggling to leave the ground, you would have seen each paddle trace out part of a circle, stop, return up, and trace part of a circle again. Had Cayley only recognized that by rotating one of the paddles *all* the way around, he could have accomplished the same result more efficiently. In other words: down, twist, and *around* – a propeller motion.

Ironically, the propeller – in the form of the rotating blades of the toy helicopter – was the very mechanism that had first intrigued Cayley. If he had just turned the helicopter sideways and attached it to his glider, he would have had an airplane.

Showing the Way

We cannot blame Cayley for failing to see what is obvious to us. We have the advantage of knowing what successful flying machines actually look like. If Cayley could see a small private plane today such as the two-seater Cessna 152, which is about the size of the machines he was contemplating, very little about it would surprise him. He had already thought of most of it. Fixed wings? ("Yes, of course," he might say.) A cross-shaped tail for stability? ("My first glider had that," he would note.) Streamlining? ("See my paper of 1809.") Three-wheeled landing gear? ("How else would you do it?") And, finally, the propeller? ("Hmm, possibly," he might venture.)

What would truly astonish Cayley is the small size and high power of today's aircraft motor. This is one of the real secrets of flight. The Cessna 152 weighs 1,700 pounds (800 kg) fully-loaded. Just one-seventh of this weight is the engine, and it generates an incredible 110 horsepower. In Cayley's day, such power was available only in the largest factories. Harnessed within a small package, the engine can push the Cessna to 130 miles per hour (200 km/h), as high as three miles (4.5 km), and, with a full tank of fuel, a distance of 800 miles (1,300 km). The Cessna is truly Cayley's dream machine.

A small Cessna – with the basic features of an airplane imagined a century and a half ago by Sir George Cayley.

Cayley's Flapping Propulsion System   **Modern Propeller**

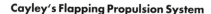

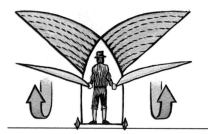

While Cayley did not anticipate the use of propellers on planes, he correctly guessed that the method that would be used to power flying machines (like the Cessna) would be *internal combustion*. Internal combustion is the burning of fuel inside an engine rather than outside. The steam engine, in which an external furnace is used to boil water to make steam, is an example of *external* combustion. Internal combustion is far more efficient than external because most of the heat of burning, which takes place inside the engine, is available to work the machinery. This translates into more power for less weight.

Though Cayley attempted to design an internal combustion engine, it was almost beyond the technology of his day. The engine had to be as carefully constructed as a clock, with precisely machined parts and automatic adjustments for the fuel and air mixture. There was also the problem of the fuel itself. Wood and coal would not do since they burned much too slowly. Something explosive was required – like gasoline, which had not yet been discovered. These and other problems delayed the invention of a practical internal combustion engine until after Cayley's death.

Cayley was ahead of his time in anticipating the problems that had to be solved before humans could really fly. And though he could not build the engine he wanted or come up with the perfect propulsive mechanism – he did lead the way.

The Flying Coachman

In 1853, during his eightieth year, Sir George Cayley had supervised while his nervous coachman mounted a boat-like carriage suspended beneath a large wing of cloth stretched over a wooden frame. Pushing off downhill, the vehicle rolled along and then lurched briefly into the air, sailing across a narrow valley at the back of Cayley's country house.

It was probably the first gliding flight in history. A human had at last flown free like a bird – or, at least, like a gliding bird. Though having hoped to achieve much more, the elderly Cayley must have been satisfied to have come so far since the simple feather helicopter of his youth. The flight was miraculous, though it lasted less than a minute.

No doubt, it seemed much longer to the pilot, for on emerging from the capsized craft, he announced to his employer, "Please, Sir George, I wish to give notice. I was hired to drive and not to fly."

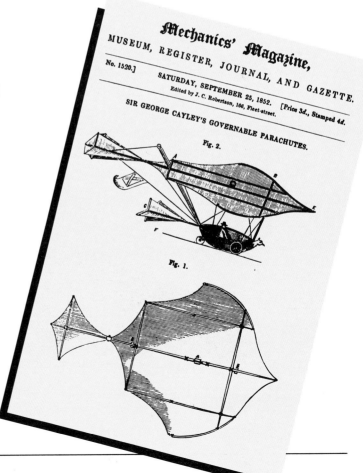

Cayley's design for a piloted glider such as his coachman probably rode. Here it is called a "governable parachute" in a British technical journal of the day.

How to Fly

A full year before Cayley's coachman took his brief glide, the balloon had just about reached perfection. Using a streamlined cigar shape (as originally suggested by Cayley) and a steam-engine-driven propeller, the Frenchman Henri Giffard was *almost* able to conquer the wind and go where he wanted. Everyone knew that a better engine (which would come within a few years) would finally do the trick.

For balloonists, engine weight was not so serious a problem, since they could always build a bigger balloon filled with more gas to lift a heavier motor. Nor was stability an issue, since a balloon is naturally stable, like a ship on a calm sea. Balloonists were also making progress in propulsion, having chosen the propeller as the most promising method of beating the wind. (Blanchard was the first to try out a propeller, in a

feeble hand-cranked version. Cayley also recommended propellers for balloons, but not, oddly enough, for winged machines.)

Awaiting an even better engine, wing enthusiasts could only practice what their wings could already do – glide. One man in particular, Otto Lilienthal in Germany, recognized that gliding would show the way to solving the control problem. It's like "learning to ride a fractious horse," said an American investigator. One method is "to get on . . . and learn by actual practice how each motion and trick may be best met," he noted. "The other is to sit on a fence and watch the beast awhile, and then retire to the house and at leisure figure out the best way of overcoming his jumps and kicks. The latter system is the safer, but the former, on the whole, turns out the larger proportion of good riders."

◄

**Otto Lilienthal (1848-1896),
first to master the art of gliding.**

Lilienthal and the American, whose name was Wilbur Wright, wanted to be good riders.

Lilienthal Learns to Ride

It is not easy to construct an airplane that is stable, much less one that turns left, right, and goes up or down as you like. ("Airplane" was the term coined during the nineteenth century to describe a powered flying machine with a fixed plane surface for wings.) The "bird suit" school of thought did not even address these problems, believing that you simply could build something with wings and then take off, trusting instinct to take care of the rest.

Lilienthal took a more scientific approach. Born in 1848, by the age of fourteen he was swooping down hillsides with homemade wings, trying to flap his way aloft. Young Lilienthal quickly concluded that he didn't know enough about the technical problems of flying and went off to engineering school. Afterward, he studied bird flight for several years, becoming the world's expert on the subject. Only in the late 1880s was he ready to resume his experiments.

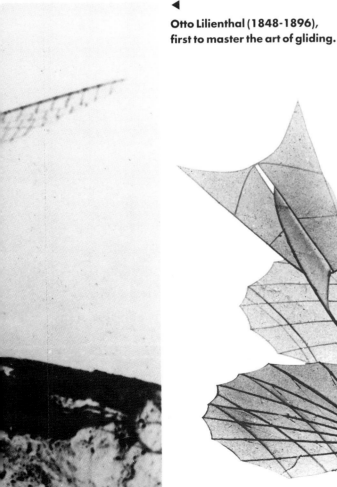

Lilienthal in flight.

Setting aside the question of propulsion, he started by focusing on control. First he tried out a simple set of wings, nothing else. These he attached to a wooden frame that he could grip. He stood on a spring-board, then bounced into the air for a brief taste of flight. Lilienthal decided that wings by themselves were un-stable, and added a tail. Much testing of various wing and tail shapes finally produced a machine that he trusted for more daring leaps. In this way, Lilienthal's glider and his flying skills advanced together.

One of the most serious problems Lilienthal faced was the tendency of any craft with wings to pitch downward in a nose dive. The opposite tendency, a sharp pitch upward and a resulting bottom-first fall, was equally dangerous.

An airplane in flight is supported by the lift pro-duced by the wings. The wings are placed close to a plane's center of gravity, or balance point. But since the plane can also pivot on this point, like a seesaw, any gust of wind or other disturbance will cause it to tip over.

To counteract this effect, Lilienthal added a tail to his glider. An airplane's tail is really two surfaces – a vertical fin and a horizontal stabilizer. The stabilizer is a small wing, angled downward into the airstream to produce a downward force on the tail end of the plane. Just as pushing down on one end of a seesaw causes the other end to go up, the downward force on the tail of a plane holds the nose up. A similar surface could also be put on the front, but it is not really needed. One stabilizer is sufficient to balance the other end of the plane.

The other part of the tail, the vertical fin, acts like the feathers on an arrow to keep the aircraft's nose pointed in the direction the plane is travelling. Without a vertical fin, the plane would tend to pivot sideways and tumble out of control.

These were hardly new discoveries made by Lilienthal. Cayley had already put vertical and hori-zontal tail surfaces on the sketch of his first winged flying machine in 1799. He was no doubt inspired by arrows (or, perhaps, by nature's examples – birds with their feathered rear stabilizers and fish with their vertical fins). It's not clear whether Lilienthal knew of Cayley's work, but he took it further nonetheless. He became the first to put these devices to a rigorous test for human flight – at the risk of his own life.

Graduating from the simple springboard to a high, artificial hill built specifically for his gliding experiments, Lilienthal became a "good rider" indeed. He made over 2,000 flights during 1891 to 1896. Not only could Lilienthal glide on a straight course down-hill, but he could change speed and direction by shift-ing his weight beneath the glider. If he shifted to the left, for example, the left wing would drop, causing a slight sideways motion which the vertical fin automati-cally corrected by making the craft pivot; the result was a neat turn to the left. Shifting his weight to the right made a rightward turn. Moving forward caused the craft to pivot down and speed up (useful when starting out). Shifting back made it slow down (good for land-ing). Like a bird learning to fly from scratch, Lilienthal was gradually discovering the ways of the air.

This modern Cessna incorpo-rates two important features first tested extensively by Otto Lilienthal – a vertical fin and a horizontal stabilizer.

Flight Test

You can discover the problems of flight control for yourself by building a model airplane with a few simple materials.

Materials
- Plastic straw
- Heavy paper (such as an index card)
- Tape
- Scissors
- Paperclip

1
Trace wing, fin, and stabilizer onto cardboard. Cut out along solid lines.

2
Cut slits as marked. Run a ball-point pen along dotted lines. Fold edges as marked.

3
Assemble as shown.

4
To fly: Set the control surfaces – ailerons, rudder, elevator – in neutral (unbent) positions. Now throw the plane. It should sail straight ahead. You may need to make minor adjustments in the position of the wings and tail to assure smooth flying. Now test each control surface by moving it up or down to see how it affects flight. (Hint: The two ailerons should be bent in opposite directions.)

Fold

Elevator

Stabilizer

Fold

Fold down

Fold

Aileron

Cut

Wing

Fold down

Fold up

Fold down

Cut

Cut

Fold

Rudder

Fin

Cut

Wing

Cut

Fold down

Aileron

Fold

Insert fin into 1½-in (4-cm) slit at top of straw.

Tape straw on top of wing. Finger hold for launching should be at bottom.

Attach paper clip to nose.

Tape stabilizer to bottom of straw. Elevator should protrude from back.

Hang High

Otto Lilienthal built a cone-shaped hill fifty feet (15 m) high as a perch for his daring gliding experiments. Today, one hundred years later, glider pilots leap from cliffs hundreds of times higher, such as those in the Sierra Nevada mountains of California. The shape and position of Owens Valley, located on one side of the Sierras, produces huge up-drafts on which hang-gliding enthusiasts can hitch a truly spectacular ride.

They often find themselves sharing the ride with birds, who, like the gliders, come equipped with broad wings that give good lift at low speeds. The most expert of the gliding birds, such as hawks, vultures, and swans, have occasionally been spotted as high as 29,000 feet (9,000 m), by jetliners – not hang gliders, who prefer to keep to the somewhat friendlier skies lower down.

But he knew that a bird does more than shift its weight in order to control where it is going. Birds twist their wings and tails in complicated ways so that they can move with precision. Lilienthal was getting ready to test such systems on a glider when, on August 9, 1896, a sudden gust of wind brought one of his routine glides to a standstill in mid flight. Instinctively, he threw his weight forward to speed up and regain lift. But it was too late. His plane crashed to the ground, critically injuring the great pilot. He died the next day.

A hang glider goes for a high flight over Colorado – a trip made possible by the briefer, but more dangerous leaps of pioneer Otto Lilienthal (inset).

Powered Hops

About this time, many other inventors were working on flying machines, but gave scarcely a thought to control. Most of them, like industrialist Hiram Maxim in England, concentrated on the motor problem. Using a fortune earned by inventing a machine gun, Maxim built the largest possible wings to lift the lightest possible steam engine. Finished in 1893, the vehicle tipped the scales at 8,000 pounds (3,600 kg), including coal, water, and a three-member crew.

Maxim's big machine actually got airborne – for a moment. He had carefully gauged the power, propeller size, wing area, and speed needed to achieve his goal of "a machine that would lift itself from the ground." And it did – reaching about knee height. Maxim was perfectly satisfied with this achievement. He didn't want to fly far; he just wanted to take a short hop.

If simply leaving the ground was the definition of flight, then Maxim had done it, and maybe others before him. As early as 1874, a young French sailor rolled down a sloping ramp and rose briefly into the air aboard a steam-powered plane built by one of his officers, Felix Du Temple (who had been inspired by reading Cayley's technical articles). A Russian named Alexander F. Mozhaiski tested a similar ramp-assisted craft in 1884. In both cases, gravity did a good part of the work of getting the machines up to flying speed. Once in the air, when the planes were no longer accelerating downhill, the engines were not enough to maintain flying speed, and the "flights" ended.

It's a good thing they did, for once in the air most inventors intended to steer their ships by a most inappropriate method – the way you would a boat.

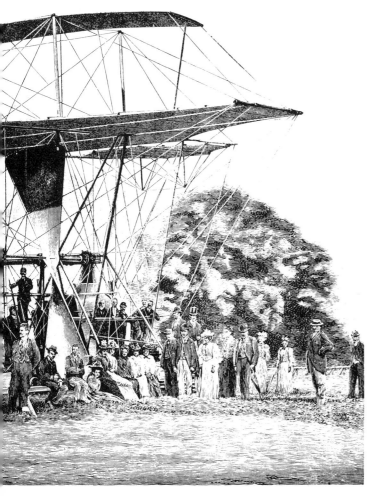

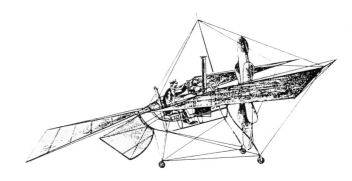

After its brief hop into the air, Hiram Maxim's monstrous flying machine (left) became a popular attraction at fairs, where it was displayed with its wings considerably shortened. Felix Du Temple (right) made a similar hop in his sportier craft a few years earlier.

Steering in Three Dimensions

In the search for the secret of flight, investigators made the most progress when they tried to look at matters in a new light. Lana made a breakthrough by seeing the similarity between water and air. The Montgolfier brothers created a flying machine after observing that smoke rises. Cayley discarded old notions about how birds fly, and discovered the way wings really work. And Lilienthal realized that the problem is not *getting* into the air but knowing what to do once you're there.

Now, someone was needed to pull all the threads together and make flight a practical reality.

In Dayton, Ohio, two brothers who built bicycles for a living had read of Lilienthal's death with a sense of sadness mingled with curiosity. Thanks to improved methods of making and printing photographs, Lilienthal's exploits had been widely publicized in magazines. His death was a shock to Wilbur and Orville Wright, who had already turned their thoughts from wheels to wings, inspired by the dramatic pictures of Otto Lilienthal aloft. Newspapers speculated on what mistake had doomed the "Bird Man," as they called Lilienthal. The Wrights, too, aimed to find out – and avoid – his fatal mistake.

Employing Lilienthal's methodical approach, the two brothers read everything they could find on the subject of flight. Their research convinced them of two things. First: gliding experiments like Lilienthal's were the most promising route to success. Second: control was the key. In the Wrights' opinion, Lilienthal hadn't paid nearly enough attention to developing precise control methods. They wouldn't make the same mistake.

The discovery that air is a fluid like water, only thinner, was the breakthrough that had set scientists on the course to solving the problem of flight. As useful as the water analogy was, it nonetheless led many investigators astray.

Most believed that steering an aircraft would be similar to steering a boat. But an airplane has to move in three dimensions – up, down, and around through the air – while a boat moves in just two – forward and to the side on the water's surface. This problem seemed easily solved by the addition of an extra control mechanism. Experimenters believed that turning an airplane to the right or left would be handled, as in a boat, with a vertical rudder. Steering up or down would be accomplished with a kind of horizontal rudder called an "elevator." Both controls could be attached to the tail: the rudder on the vertical fin, the elevator on the horizontal stabilizer. There was already a craft like this, the submarine, that used the very same controls to operate in three dimensions under the sea.

The trouble is that ships and submarines do not have wings. As Wilbur Wright pointed out, a boat stays afloat no matter what its speed, but an airplane must keep moving at all costs. Any attempt to steer must not

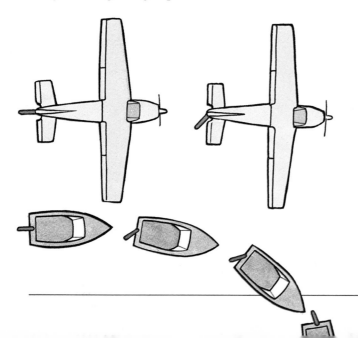

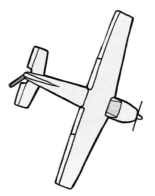

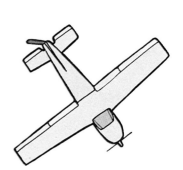

A boat's rudder produces a neat turn, while the rudder of an airplane, if used alone, produces a skid. If the rudder is applied long enough, the plane will go into a corkscrew dive and crash.

upset the critical forces that hold the plane up. An airplane therefore has a totally different set of problems from a boat.

For example, a pilot who adjusts the elevator up too high, in an attempt to make the plane go up, can easily plunge to the ground. Why? An up-turned elevator pushes the nose up and the tail down; this increases the tilt of the wings, and increases lift – up to a point. Beyond a certain angle, the wings lose their ability to produce *any* lift (a condition called a "stall"), and the plane starts to fall.

Using the rudder to make a turn can be equally dangerous. A pilot who tries to turn left by holding the rudder hard to the left – as you would a boat – will find the aircraft skidding sideways and, eventually, into a fatal downward spiral. It's not what the average investigator would predict.

But the Wrights were hardly average investigators. One solution seemed obvious to them. As bicycle specialists, they knew that turns are made by banking – leaning into a turn. They had even observed that to start a turn, say, to the left, a cyclist must first point the handlebars a little to the right; the machine responds by inclining to the left. The rider then leans left while turning the handlebars left, and makes the turn.

This exercise in analyzing what everyone can do but hardly anyone understands taught the brothers that steering is an art of surprises. The lesson served them well in learning to ride their "fractious" new horse – the Wright glider.

The Perfect Turn

From the start, Wilbur and Orville equipped their glider with a control system. The first mechanism they wished to try out was a method of banking, or turning.

On a bicycle, you bank by moving the center of gravity. This is what Lilienthal did to turn his glider; he shifted the center of gravity by leaning in the direction he wanted to turn. The glider responded by rolling slightly, like a chicken on a rotisserie, and made the turn naturally.

The Wrights wished to achieve the same roll maneuver by purely mechanical means. They believed that the way to do it was to make one wing produce more lift than the other. The increased lift on one wing would cause that side to rise and the other to fall, resulting in a roll.

Banked turn

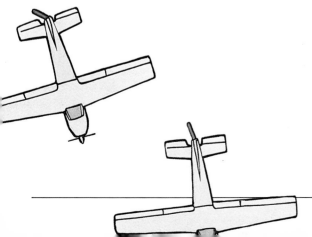

The Wright brothers knew that if they could get an airplane to bank, it would make a turn naturally – the way a bicycle does.

How to Fly a Plane

Three controls are needed to fly a plane – one for the rudder, one to twist the wings (or operate the ailerons), and one for the elevator. This raises the question of how to do it with just two hands.

On the original Wright flyer, the brothers solved the problem by using their hips. They rigged a cradle that the pilot, who was lying face-down, could operate to make turns. The cradle was linked to the wings and vertical rudder. When the pilot moved his hips to the left, the cradle mechanism would simultaneously twist the wings and turn the rudder for a left turn. A hand-controlled lever worked the elevator for up and down maneuvering.

Airplanes soon evolved with the pilot sitting rather than lying. The rudder is no longer connected to the action of the ailerons (which produce the same effect as wing twisting). The new controls use foot pedals to work the rudder, and a double-acting lever, or stick, to operate both the elevator and the ailerons.

With this method, pushing the stick forward moves the elevator down and causes the nose to pitch down; pulling the stick back has the opposite effect. Moving the stick left or right causes the ailerons to hinge into position for a left or right roll. At the start of a roll, the pilot points the nose into the turn by depressing the proper rudder pedal, which controls *yaw*, the side-to-side pivoting of the plane. The pilot depresses the left rudder for a left turn, the right rudder for a right turn.

Nowadays, the stick on many airplanes resembles a steering wheel that can be turned left and right, or pushed in and out. The rudder pedals are unchanged.

Not to be forgotten is the throttle, which controls the amount of fuel going to the engine and therefore the amount of propulsion. The throttle is the true altitude control. Using the throttle to increase propulsion causes the plane to speed up; this causes the wings to produce more lift, and makes the plane rise.

Working the throttle is something for your right hand to do while your left hand works the ailerons and elevator, and your feet are busy with the rudder. It takes practice.

Cockpit of a small jet airliner. Two sets of controls are provided so that the plane can be flown from either seat.

Control Column ("Stick") Throttle Rudder Pedals

An airplane rotates around three axes – called pitch, yaw, and roll.

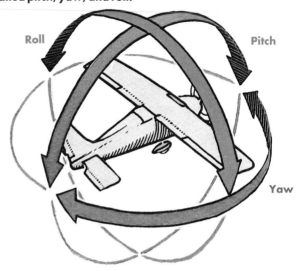

Roll Pitch Yaw

Pitch
Elevator up: pitch up.   Elevator down: pitch down.

Yaw
Rudder left: yaw left.   Rudder right: yaw right.

Roll
Left aileron down, right aileron up: roll right.   Left aileron up, right aileron down: roll left.

Seeking a reliably windy location for their experiments, the Wrights travelled to Kitty Hawk on the coast of North Carolina, which had been recommended by the U.S. Weather Bureau. There, in the early fall of 1900, the brothers tested an idea that had been inspired, oddly enough, by an empty box.

During a slow day at the shop back in Ohio, Wilbur absent-mindedly twisted an inner-tube box. He noticed that the box had the same shape as the double-winged biplane that he and Orville were working on. (A biplane has two wings stacked one atop the other and connected by *struts*, or supports, to form a box-like shape.) One side of the box angled up, the other down. Wilbur realized that if this were a set of wings, one side would produce more lift, the other less, and the craft would roll in the direction of the down-turned wing. If the box could be easily twisted, then perhaps wires attached to the wing tips on the biplane could accomplish the same thing. They rigged their first glider for wing twisting (or "wing warping" as it came to be called), and it worked. But the turn was far from smooth.

During the following two years, more experiments at Kitty Hawk confirmed that wing twisting had a dangerous and unexpected side effect. While one wing did produce more lift than the other, it also generated more *drag*, or air resistance, causing that wing to slow down. Meanwhile, the other wing was producing less drag and was trying to speed up. The result was that the plane tended to pivot in the *opposite* direction to the one in which they wished to turn. They would steer to the left, but would point right instead! The brothers had more than one crash due to this problem, and almost gave up because of the apparent hopelessness of steering in the air.

Wilbur and Orville knew that somehow they needed to apply a force on the plane to counteract the tendency to swerve in the direction opposite the desired turn. At the outset, they had considered a vertical rudder unnecessary and dangerous as a means of turning. But after extensive experimenting they found that the undesirable skid caused by a rudder was actually perfect for canceling out the unwanted swerving. With a bank produced by wing twisting, and a gentle skid supplied by the rudder, they could execute perfect turns. To complete their controls, the Wrights used an elevator on the front of their plane for limited maneuvering up and down.

Today's airplanes use the same basic flight control system developed by the Wrights, with a few exceptions. The elevator now is usually mounted on the tail, and wing twisting has been replaced by the more efficient action of "ailerons," which are hinged flaps on the wings that move in opposite directions to produce unbalanced lift and a roll.

Twist up: more lift.

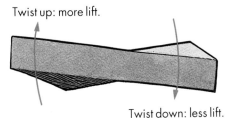

Twist down: less lift.

The ease with which a box can be twisted (above) inspired the Wright brothers to try the same thing on their glider, which had a box-like shape (right). An upward twisted wing produced more lift, a downward twisted wing less. This caused the craft to roll and make a turn. The red lines show the position of the control wires.

Airships

While the Wright brothers were having trouble with aerial turns at Kitty Hawk, Alberto Santos-Dumont of Brazil was flying rings around the Eiffel Tower in Paris. Santos-Dumont was a pioneer of "dirigibles," or steerable balloons. He was the first to make air travel practical, using his ship for trips to his favorite night spot.

More practical still were the rigid airships being developed about the same time by Count von Zeppelin in Germany. Employing an aluminum framework, these massive monsters were eventually able to carry more than seventy passengers, along with elaborate accommodations. Sixty years ago "zeppelins," as they were called, seemed destined to dominate the market for long distance air travel. But the explosion of the largest of them, the *Hindenburg*, in 1937, caused the public to turn against balloon travel, and gave hydrogen, the inflammable lifting gas that filled the *Hindenburg*, a bad name.

As enormous as they are, modern blimps are tiny compared to the great zeppelins of the 1930s. Blimps lack the rigid skeleton that allowed the "zeps" to be so big. They also use helium, a less efficient but safer lifting gas than hydrogen.

The inventors of powered flight: Orville Wright (1871-1948), with the moustache, and his brother Wilbur (1867-1912).

At the end of 1902, over a thousand flights and countless hours of tinkering by the Wright brothers had produced a truly agile machine, and the experienced pilots to fly it. "When once a machine is under proper control under all conditions," Wilbur had predicted to his father in 1900, "the motor problem will be quickly solved." Next came the solving of that problem.

The Final Pieces of the Puzzle

At the start of the twentieth century, one of the most famous scientists in America tried to put a damper on public excitement about flying machines. "The example of the bird does not prove that man can fly," declared Professor Simon Newcomb of the United States Naval Observatory. Newcomb gave a mathematical proof showing that a successful flyer could be no larger than a watch and carry nothing more than a button.

About the same time, Wilbur and Orville Wright were doing some mathematics of their own. The brothers calculated that an engine of at least eight to nine

horsepower, weighing no more than 200 pounds (90 kg), would be sufficient to lift their flying machine along with its pilot – buttons and all.

The brothers were counting on the rapidly developing automobile industry to meet their need for a lightweight and powerful engine. By 1902, when the Wrights sent inquiries to a number of motor manufacturers, the latest sensation on American roads was the curved-dash Oldsmobile. This was a two-passenger buggy powered by a four horsepower, internal combustion gasoline engine. Engines had come a long way since James Watt's bulky, coal-fired steam models, but not quite far enough. Auto makers designed their motors to be rugged, not light, and they expressed no interest in building one to the Wrights' specifications. The brothers would have to build it themselves.

Under necessity, the bicycle specialists became superb engine mechanics. In just six weeks, they completed a motor that produced about one-half more power than their airplane would need.

Then came the problem of a propeller. Broad, fan-shaped propellers had been used successfully on steamships for years, and most balloon and airplane experimenters adopted similar shapes. But these proved oddly inefficient. The Wrights recognized that a propeller is simply a rotating wing – something that Cayley knew but that practically no one else had followed up on. The brothers studied the data they had gathered in designing the wings for their glider and used the information to design a very efficient propeller that was entirely suited for air, rather than water.

Exulting in their success at tackling the problems of flight, one after another, Orville wrote to a friend, "Isn't it astonishing that all these secrets have been preserved for so many years just so that we could discover them?"

Engines of the Air

An engine is a machine for converting energy into motion. A lawnmower engine, for example, takes the energy released from burning gasoline and harnesses it to push a piston. The piston turns a shaft, which wields a grass-cutting blade around and around. The same engine can be hooked up to blow snow, pump water, turn wheels, or work a propeller.

The purpose of a propeller is to move air. An ordinary house fan (which is simply a propeller connected to an electric engine) pulls air from behind the fan and moves it forward, producing a pleasant breeze. If you place the fan on a small cart, the cart will move in the direction opposite the breeze. A propeller-driven plane works the same way; it pushes air in one direction and moves in the opposite direction in reaction.

A more powerful method of moving air is to do away with the piston and propeller, and let heated air do the pushing. This is the principle of the jet. The simplest kind of jet engine is an open tube, where air enters at the front, combines with a sprayed mist of fuel, ignites, and then expands out the back, travelling at a much higher speed than when it entered at the front.

Close the front of a jet engine and introduce an "oxidizer" (which is needed to burn the fuel) and you have a rocket. Rockets travel even faster than jets, and can be used where air is not available, as in space.

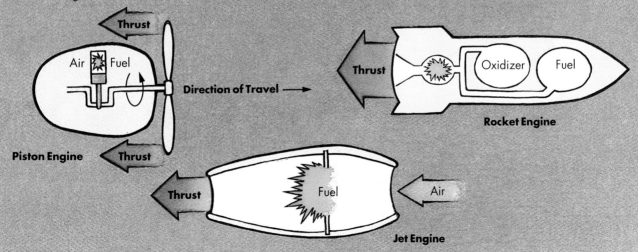

Thrust

Air · Fuel

Direction of Travel →

Piston Engine · Thrust

Thrust · Oxidizer · Fuel

Rocket Engine

Thrust · Fuel · Air

Jet Engine

How to Fly

On December 17, 1903 – 120 years after humans first tasted flight in a simple balloon over Paris – Orville Wright climbed onto the complicated machine that he and his brother had so diligently perfected. He adjusted the motor, checked the controls, and released the brake. He became, by his own description, the first person to take off in a machine raising itself into the air by its own power, sailing forward without reduction of speed, and landing at a point as high as that from which it started.

You could jump nearly as high as Orville went on that outing at Kitty Hawk. And you could easily run faster and farther. Nevertheless, it was the first controlled, sustained, and powered flight in history, and the brothers would make three more, alternating at the controls, before the short winter day was over.

Oddly enough, as soon as the accomplishment became known, others stepped forward to claim they had done as much, usually in secret and without witnesses. But the Wright brothers had done it in the open and needed no magic. What they did need was the insights of Lana, the Montgolfier brothers, Charles, Cayley, Lilienthal, and countless others who had pursued the dream of flight with open eyes and inquiring minds.

Part of the Wrights' genius is that they created a method of flying that is basically unchanged today. As we have seen, others had already connected a propeller to an engine and made powered hops. Eventually, still others would have probably made sustained flights and solved the control problem after a few crashes. But the resourceful Wrights were able to put it all together in one package – to discover all the secrets – before taking off and showing the world exactly how to fly.

The air age begins. With Wilbur looking on, Orville Wright takes off on history's first controlled, sustained, and powered flight, December 17, 1903.

Onward and Upward

No sooner had the Wright brothers shown the proper way to fly than experimenters began working to improve every aspect of the new art. The process continues even to this day, as engineers devise more efficient wings, structures, and engines in the quest to fly ever faster, higher, and farther.

In this graph, airplanes that go higher appear higher on the page. Those that go faster appear farther to the right. *Mach 1* on the velocity scale is the speed of sound, equal to 660 miles per hour (1,060 km/hr) at the altitude where *supersonic* (faster-than-sound) jets usually fly. *Mach 2* is twice the speed of sound, etc.

The fastest, highest flying machines of all are those powered by rockets (far right graph).

World War I ace Eddie Ricken-backer with his Spad fighter.

Charles Lindbergh and the *Spirit of St. Louis,* first to fly solo across the Atlantic.

Chuck Yeager and the X-1, first to break the sound barrier.

P-51, top dogfighter of World War II.

▶ **F-14,** 1972

◀ **Concorde,** 1969

✈ **X-1,** 1947

✈ **P-51,** 1943

✈ **DC-3,** 1936

Spad, 1916

Spirit of St. Louis, 1927

Altitude in miles (km)

▮ **Wright Flyer,** 1903

25 (40)
20 (32)
15 (24)
10 (16)
5 (8)

DC-3, one of the first and most successful modern airliners.

Concorde supersonic airliner.

Speed in miles (km) per hour

| 500 (800) | Mach 1 | 1,000 (1,600) | Mach 2 | 1,500 (2,400) |

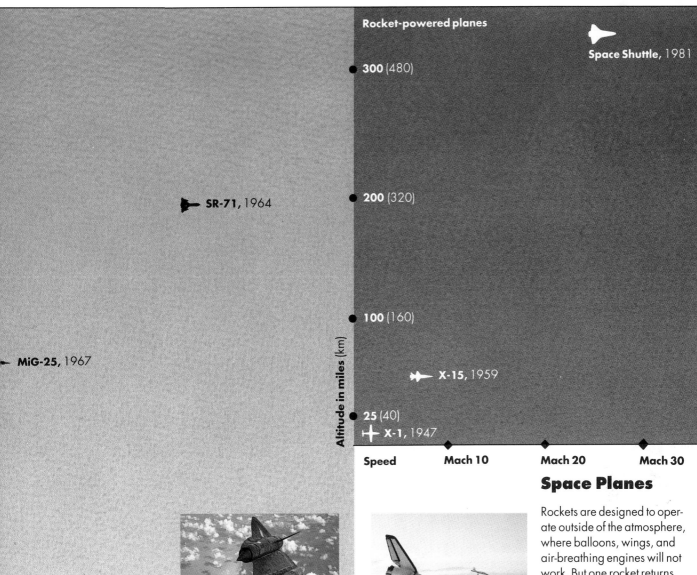

Rocket-powered planes

Space Shuttle, 1981

300 (480)

200 (320)

SR-71, 1964

100 (160)

MiG-25, 1967

X-15, 1959

25 (40)

X-1, 1947

Altitude in miles (km)

Speed   **Mach 10**   **Mach 20**   **Mach 30**

Space Planes

Rockets are designed to operate outside of the atmosphere, where balloons, wings, and air-breathing engines will not work. But one rocket returns from space in much the same way that Otto Lilienthal leaped from his artificial hill.

The space shuttle uses its wings and control surfaces to glide from the top of the atmosphere to a pinpoint landing on the ground. Is the shuttle a true space plane? Not really, since during launching it relies on rocket thrust and a steep flight path that generates little lift. Future launchers, however, are being designed to use advanced air-breathing engines plus lift generated by wings for an efficient, airplane-like ride into space.

SR-71, the current speed and altitude champ.

Space Shuttle gliding to earth.

The speedy, high-flying Soviet MiG-25.

U.S. Navy pilot takes off in an F-14.

X-15, about to be launched from a bomber.

2,500 Mach 4
(4,0Q0)

Index

Photo Credits

Front cover © John Heiney **Back cover** © Jim Goodwin/Photo Researchers **1** National Air and Space Museum **2-3** © Russell Munson **4-5** © Jordan Coonrad; (inset) © Art Davis **6** National Air and Space Museum **6-7** (both) NASA **8** Granger Collection, NY **9** (left) Granger Collection, NY; (right) Bettmann Archive **10** (left) courtesy of J. Kim Vandiver and Harold E. Edgerton; (right) Granger Collection, NY **12** Bettmann Archive **14** (left) Bettmann Archive; (right) © Ralph Starkweather/West Light **15** (left) © Brent Bear/West Light; (right) © M. Serraillier/Photo Researchers **16** from *Jean-Pierre Blanchard, Physicien-Aeronaute* by Leon Coutil, Evreux, France, 1911 **16-17** © Douglas Faulkner/Photo Researchers; (inset) Granger Collection, NY **18** Granger Collection, NY **19** WGBH photo, feather courtesy of Peter Vickery, Wildlife Dept., University of Maine **22** Granger Collection, NY **23** (Stealth bomber) Northrop; (eagle) © Thomas Kitchin/Tom Stack; (dragonfly) © Stephen Dalton/Photo Researchers; (Boeing 727) © George Hall/Check Six; (biplane) © Russell Munson; (home-built) © Jordan Coonrad; (paraglider) © John Heiney **24-25** (all) Granger Collection, NY **26** Granger Collection, NY **27** © Mike Smith **28** © Frans Lanting/Minden Pictures **29** © Kennan Ward **30** © Russell Munson **31** from *Sir George Cayley's Aeronautics* by Charles H. Gibbs-Smith, London, 1962 **32** (inset) Bettmann Archive **32-33** (both) National Air and Space Museum **34** © Russell Munson **36** (left) © John Heiney; (inset) Granger Collection, NY **36-37** from *Transportation*, ed. Jim Harter, New York, 1984; (right) from *Aviation* by Charles H. Gibbs-Smith, London, 1970 **39** © Bernard Vallet/Photo Researchers **40** © David O. Hill/Photo Researchers **41** National Air and Space Museum **42** (left) George Hall/Check Six; (right) National Air and Space Museum **44-45** National Air and Space Museum **46-47** (Spad) Culver Pictures; (Spirit of St. Louis) Bettmann Archive; (DC-3) © Russell Munson; (P-51) © Russell Munson; (X-1) Bettmann Archive; (Concorde) © Brenda Lewison/The Stock Market; (MiG-25) Wide World Photos; (F-14) © George Hall/Check Six; (SR-71) Lockheed; (X-15) NASA; (Space Shuttle) NASA

All illustrations by Brian Lies.

Produced by WGBH Publishing and Design.